I0814262

WHY LEARN CODING?

BY SAMANTHA S. BELL

CONTENT CONSULTANT
DAVID A. LASH
ASSOCIATE PROFESSOR OF COMPUTER SCIENCE
AURORA UNIVERSITY

Kids Core
An Imprint of Abdo Publishing
abdobooks.com

abdobooks.com

Published by Abdo Publishing, a division of ABDO, PO Box 398166, Minneapolis, Minnesota 55439.

Printed in the United States of America, North Mankato, Minnesota.
102023
012024

Cover Photo: He Luqi/Qianlong.com/VCG/Getty Images
Interior Photos: Sjoerd van der Wal/Getty Images News/Getty Images, 4–5; Andrej Sokolow/picture alliance/Getty Images, 7; Yuriy Rumar/Shutterstock Images, 9; Shutterstock Images, 10, 12–13, 14, 17, 24, 26, 28 (top), 29 (top); Sebastian Willnow/picture alliance/Getty Images, 18, 29 (bottom); Jacob Lund/Shutterstock Images, 20–21, 28 (bottom); Red Line Editorial, 23

Editor: Katharine Hale
Series Designer: Katharine Hale

Library of Congress Control Number: 2023939626

Publisher's Cataloging-in-Publication Data

Names: Bell, Samantha S., author.
Title: Why learn coding? / by Samantha S. Bell
Description: Minneapolis, Minnesota: Abdo Publishing, 2024 | Series: Let's code! | Includes online resources and index.
Identifiers: ISBN 9781098292805 (lib. bdg.) | ISBN 9798384910749 (ebook)
Subjects: LCSH: Coding theory--Juvenile literature. | Computer programming--Juvenile literature. | Computers and children--Juvenile literature. | Computer programming--Study and teaching--Juvenile literature.
Classification: DDC 005.1--dc23

CONTENTS

Technology is a major part of modern cars.

CHAPTER 1

THE POWER OF CODING

Katie slid into the back seat of the car beside the stack of library books. Her older brother Alex had already claimed the front seat. Katie buckled her seat belt.

"All set?" Mom asked. "We have a lot of errands to do today." She started the car.

"Where are we going first?" Katie asked.

"Can you guess?" Mom answered as they turned the corner.

"The library!" Katie said, looking at the books beside her.

"Hey, look at that car ahead of us," Alex said. "I think it's a self-driving car!"

"A what?" Katie asked.

"It's a car that can drive itself," Alex replied.

Katie laughed. "There's no such thing!"

"I just read about them," Alex said. "Some cars can do a lot of things to help drivers. They can take over most of the steering and braking. Some cars can even change lanes." Alex pointed at the car as they passed it. "And some don't need drivers at all!"

Sensors work to keep self-driving cars from running into other vehicles.

Katie took a closer look. No one was in the driver's seat! "Why doesn't it crash?" she asked.

"See that big sensor on the top?" Alex asked. "It keeps track of the car's surroundings.

There are sensors on the sides, front, and back of the car too."

"How do they work?" Katie asked.

"The sensors detect other cars nearby," Alex explained. "It takes a lot of computer code to make those sensors work!"

"One day, maybe everyone's car will be able to drive by itself," Mom said.

Artificial Intelligence

Self-driving cars use artificial intelligence (AI) to make decisions. Machine learning is a type of AI. The machines can do tasks and solve problems. They can learn, think, and act intelligently. Programmers are still finding ways for AI to improve and expand.

Technologies such as self-driving cars rely on programs written in code.

"Maybe our car could even return my library books," Katie said, laughing.

Technology for the Future

Several companies are making self-driving cars like the one Alex saw. These cars can require more than 250 million lines of code. Code is used to write programs. Programs give instructions to a computer. The computer can then carry out a task.

People who write code are called programmers. They have important roles in many industries.

All modern cars require code to work correctly. Code creates the car's **software**. This software is needed for advanced safety features.

Developing self-driving cars is just one way engineers and programmers use coding. Coding is also an important part of most industries today. Some examples include health care, education, engineering, and **finance**. Technology depends on coding. If people know how to code, they have many different job opportunities.

Coding is not just for people with programming jobs. It is a good skill for everyone to learn. Coding helps people sharpen other skills, such as creativity and **critical thinking**. People can use these skills in school, work, and other activities.

Further Evidence

Look at the website below. Does the video give any new evidence to support Chapter One?

Self-Driving Cars

abdocorelibrary.com/why-learn-coding

Learning to code can make people better at solving problems.

CHAPTER 2

LEARNING THROUGH CODING

When people learn to code, they learn other skills too. For example, coding helps people improve their problem-solving skills. When someone writes code, they follow the same steps used to solve other kinds of problems.

Computer code does not always work on the first try. Programmers learn not to give up.

First, the person identifies the problem. Then they plan how to solve it. If the plan does not work, they try something else. Programmers think about coding the same way. They think about the problem, plan their code, write their program, and then see if it works. If it doesn't, they change the code and try again.

Coding helps people in other ways. It can help people break problems down into smaller parts. Sometimes a problem is big or complicated. It can be difficult to solve. Breaking it down into smaller parts can make it easier. This is called computational thinking. Programmers often do this when they are writing code.

Programmers also learn to be persistent. They keep working with the code until they get it right. They learn to be creative as they solve problems with their programs and make them work.

Coding Careers

People who know how to code have many different career options. They might choose to create and maintain websites. They could make video games or mobile apps. They might keep computer systems safe from **cyberthreats**. Some programmers create software for computers and **smart** devices.

Coding for Fun

Not everyone learns to code for career reasons. Some people learn because they enjoy coding. They might build websites, develop games, or create mobile apps in their free time. Coding can be a lot of fun. It can help exercise the brain. Also, working through a coding problem can feel rewarding.

Programmers develop software to help keep computers safe from hackers.

Not only tech companies rely on technology. Programmers can also find work in many other industries. Programmers may work with banking systems, including online banking and security.

Field robots can pull weeds on farms. Programmers coded the programs that run the robots.

They also create the systems people use to send texts and make phone calls. Coding is also important in agriculture. Programmers create equipment that helps farmers with their crops and animals. Anytime a computer system is needed, coding is involved.

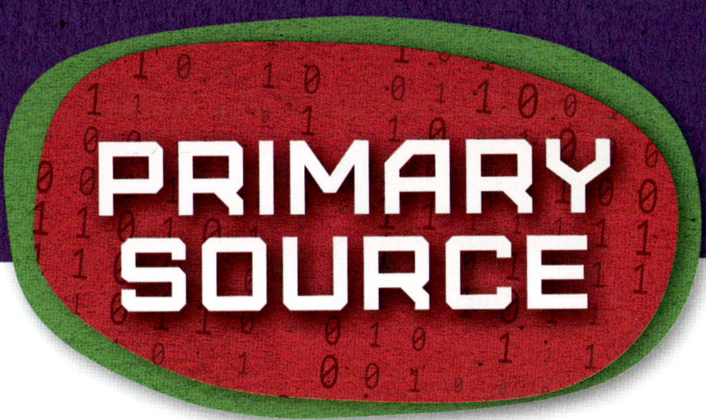

Professor Marina Umaschi Bers explained how learning to code can help young people:

> [Children] learn how to manage frustration and find a solution, rather than giving up when things get challenging. They develop strategies for **debugging** their projects. They learn to [work together] with others and they grow proud of their hard work.

Source: "How Coding Provides Skills That Can Help Children Cope with Distress." *Columbia Engineering*, n.d., bootcamp.cvn.columbia.edu. Accessed 11 May 2023.

Comparing Texts

Does the quote support the information in this chapter? Or does it give a different perspective?

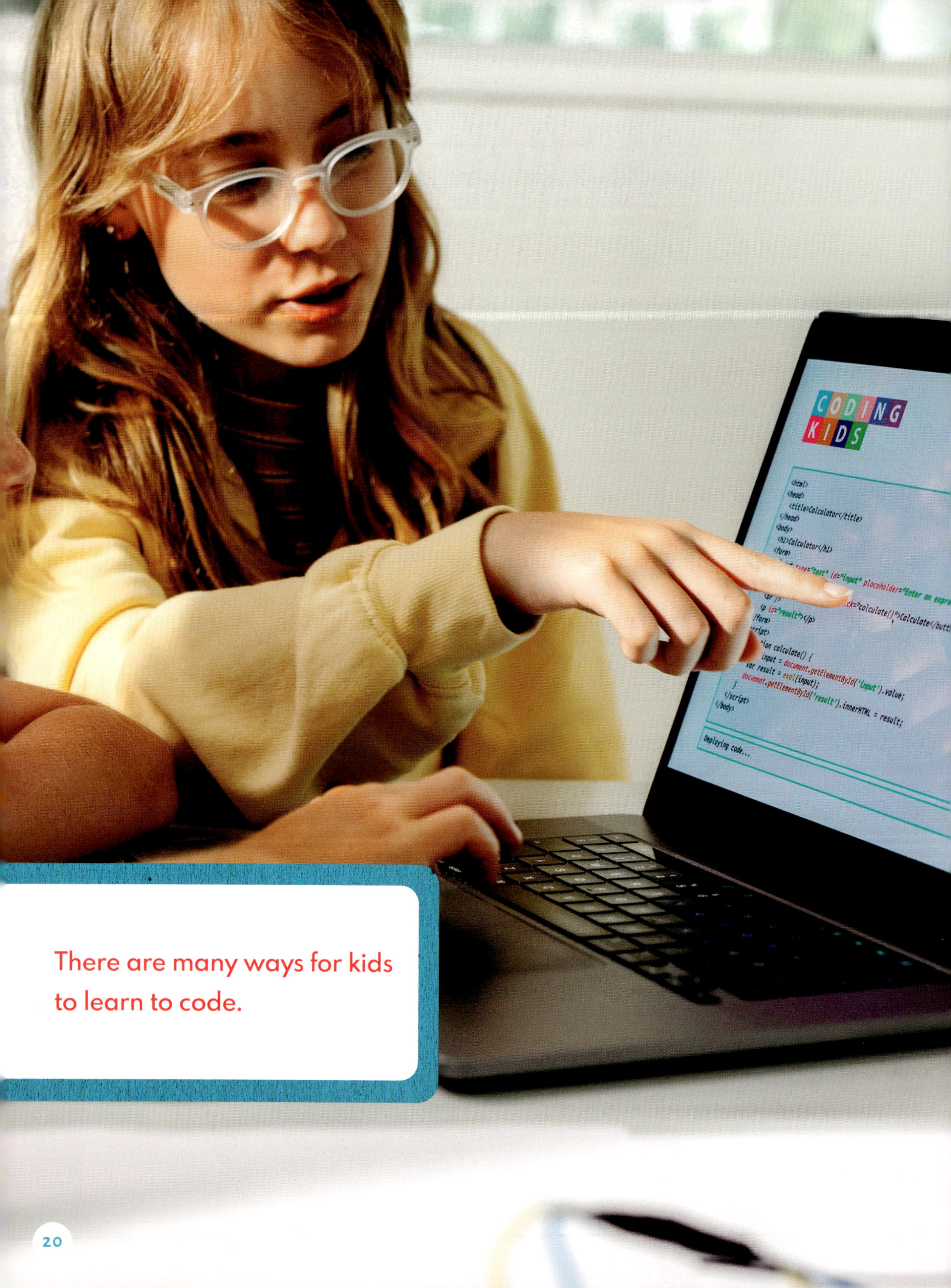

There are many ways for kids to learn to code.

GET CODING

Coding may seem complicated. But people can learn to code at any age. Online coding programs are designed for children as young as five years old. Children can use interactive stories and games to solve problems and design projects.

Some people may think they are too old to learn to code. But many older adults learn coding too. Sometimes they want to start a new career. Others may be looking for a new challenge. Some start coding as a hobby.

Learning to Code

There are different ways to learn to code. Many programmers go to college to study computer science. They earn a college degree. Some employers want programmers who have a degree. People with degrees learn more than just how to code. They also study other topics. These include advanced math concepts. This knowledge helps with advanced programming.

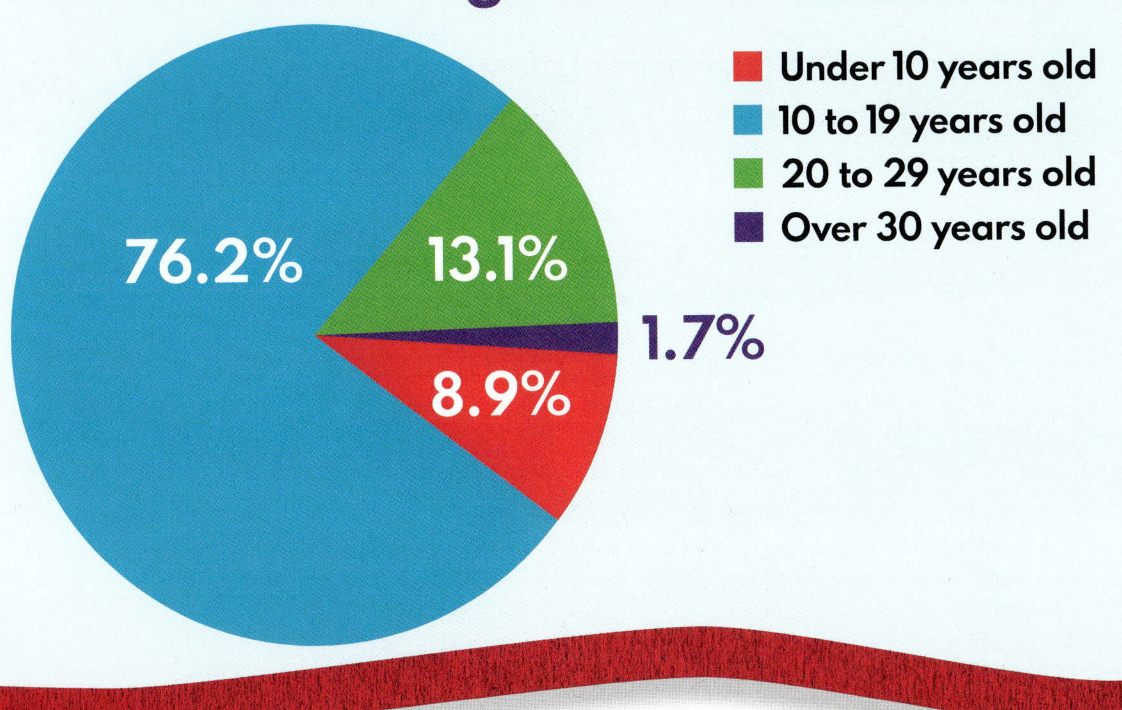

Stack Overflow is a website where programmers can share information with each other. In 2020, it sent a survey to its users. It asked when programmers started learning to code.

But there are other ways to learn coding. Some people attend coding boot camps. Boot camps take less time than a college degree. They are shorter courses that usually last only a few months. Boot camps are also less expensive. They focus on practical skills.

Block-based programs help beginners learn how coding works

Students quickly learn the skills they need to start a career in coding.

Some people teach themselves coding. They may take online courses. They may use videos and books to create their own coding projects. They learn more with each project they complete.

The easiest way to start coding is with programs that use blocks of code. These blocks are computer commands that are grouped together. Beginners do not have to write the code. Instead, they arrange the blocks in a certain order. This tells the program what to do. People learn how code works without having to write it themselves.

Some people start coding by writing in a traditional **programming language**. These languages are more difficult to learn.

Scratch It Out

One of the most popular programs for learning block-based coding is called Scratch. Users can create a character or an object, called a sprite. Then they use block coding to move the sprite, change its size, or make it say something. Users can also create their own games in Scratch.

Learning to code helps people in school, work, and more.

But people can do much more with them. They can build more advanced programs.

Learning to code isn't just about writing computer programs. People who code improve their problem-solving skills. They also get better at critical thinking. They learn to be creative and work through failure. These are important life skills.

Explore Online

Watch the video on the website below. Does it give any new information that wasn't in Chapter Three?

How to Make a Video Game

abdocorelibrary.com/why-learn-coding

CODING DATA

Coding teaches people skills such as problem-solving, critical thinking, and creativity.

People can learn to code at any age.

People can learn to code in college, in boot camps, or on their own.

Programmers can find jobs in many different types of industries.

Glossary

critical thinking
using observations and evidence to come to a conclusion about a situation or problem

cyberthreat
anything that has the potential to cause serious harm to a computer system

debugging
the process of removing errors, known as bugs, from code

finance
the businesses and activities related to handling money

programming language
a vocabulary and set of rules for writing computer programs

smart
when referring to devices, describes appliances and other household items that can connect to the internet and other smart devices

software
all of the programs that give a computer instructions

Online Resources

To learn more about coding, visit our free resource websites below.

Visit **abdocorelibrary.com** or scan this QR code for free Common Core resources for teachers and students, including vetted activities, multimedia, and booklinks, for deeper subject comprehension.

Visit **abdobooklinks.com** or scan this QR code for free additional online weblinks for further learning. These links are routinely monitored and updated to provide the most current information available.

Learn More

González, Echo Elise. *Logic in Coding.* World Book, 2021.

Hamilton, S. L. *Tesla.* Abdo, 2023.

Normandeau, Sheryl. *Become a Web Developer.* ReferencePoint Press, 2022.

Index

About the Author

Samantha Bell lives in the foothills of the Blue Ridge Mountains with her family and lots of cats. She has written more than 150 nonfiction books for students. She learned HTML through building websites and wants to learn more programming languages.